The Worm Ladies of Charlestown, Inc.
Charlestown, Rhode Island

www.wormladies.com

ISBN:1481862421
ISBN-13: 978-1481862424

# What is a red wiggler?

A red wiggler is a special worm that is used for composting.

# What is compost?

Compost is food scraps, yard waste, paper, cardboard, dryer lint, and more. These items are put into a compost bin to decay (get yucky). Making compost is a form of recycling.

# What is recycling?

Recycling is when you turn one thing into another thing instead of throwing it away. When you take an egg carton and use it to plant seeds, you are recycling the egg carton. You will be turning your garbage into castings (worm poop) that are used to help plants grow.

# What is vermicomposting?

When you add red wigglers to your compost bin, you are doing what is called vermicomposting.

# Where do red wigglers live?

Red wiggler worms live in compost bins
that can be kept indoors or outdoors. They like to eat
manure, rotten fruit, vegetables, and yard waste.
Red wiggler worms cannot be found in your backyard.

# Do worms hide?

Red wigglers hide when they feel vibrations in their worm bin. Hiding is a way of protecting themselves from animals that eat them. Birds, skunks and moles eat worms.

Red wigglers also hide from the sun. They do not like the light. The light will dry their skin and they won't be able to breathe.

# How does a red wiggler move?

The red wiggler's body is a tube with many, many segments. Bristles that stick out of the bottom of the worm help it move through the compost. Like humans, worms also have muscles that help them move.

# Does a worm have eyes?

Red wigglers do not have eyes or ears, but they do have a mouth. Worms move by a sense of touch.

# Which end of the worm is the head?

The head is on the front end and is more pointed than the back end.

# Do red wigglers have a top side?

Yes.  The bottom side is a little paler than the top.

# How do you tell the difference between girl worms and boy worms?

You can't tell the difference.  A red wiggler is both a boy and a girl.

# What is the fat lump on the worm?

The fat lump is called the Clitellum.  The red wiggler uses the clitellum to form a cocoon.

# What is a cocoon?

A cocoon is formed when worms mate. The cocoon is made of thick slime and holds the worm's eggs.

# What hatches out of a cocoon?

Baby worms, called threads, hatch out of the eggs in the cocoon.

# How many children can red wigglers have?

Red wigglers can have lots and lots of children and grandchildren and great -grandchildren.

# What does a worm look like inside?

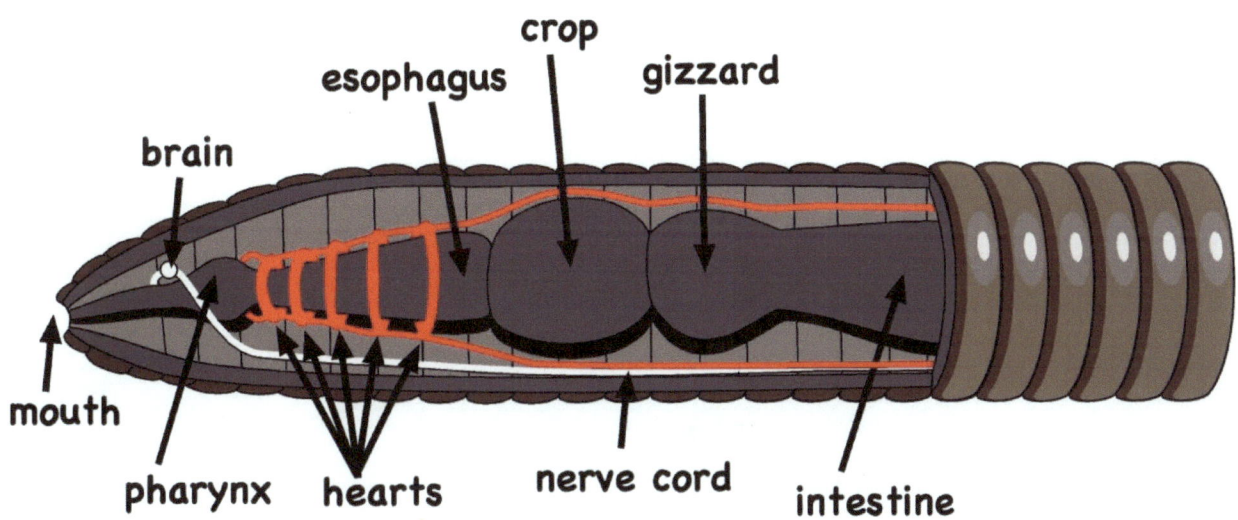

# How many hearts does a worm have?

A worm has five hearts, which pump blood through its blood vessels. The blood carries oxygen to all the cells in the body.

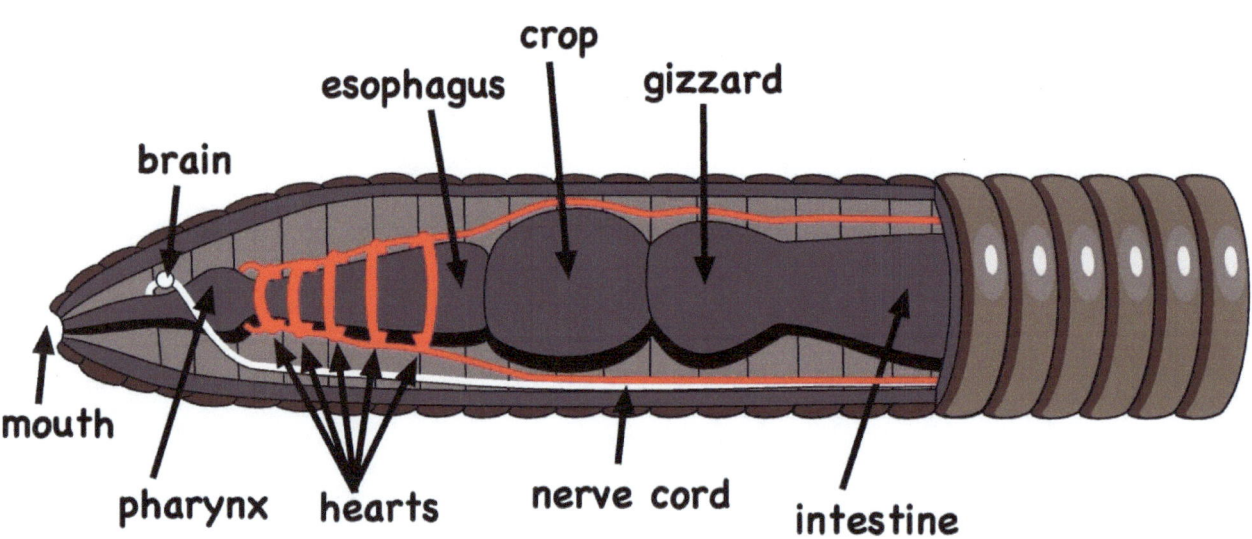

# Where does their food go?

The food enters through the mouth and moves through the pharynx and the esophagus.

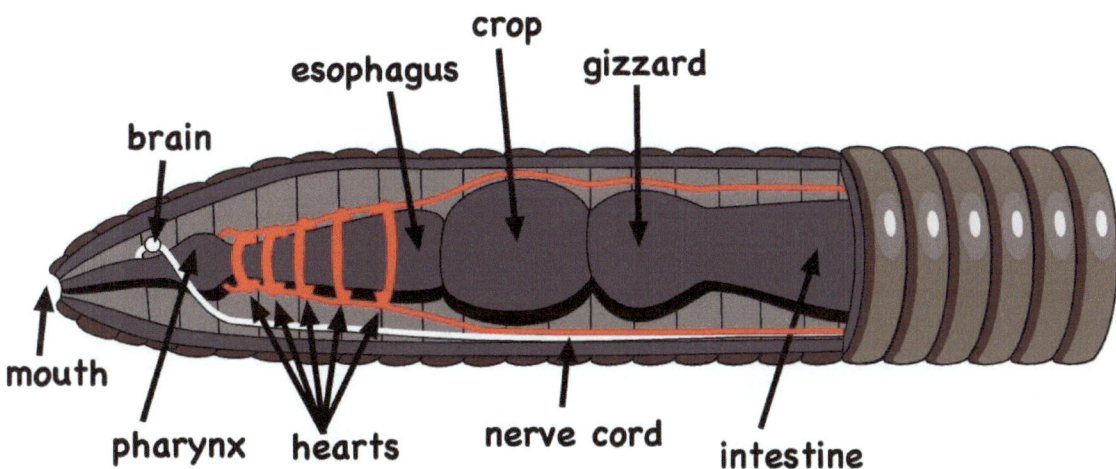

# What does the gizzard do?

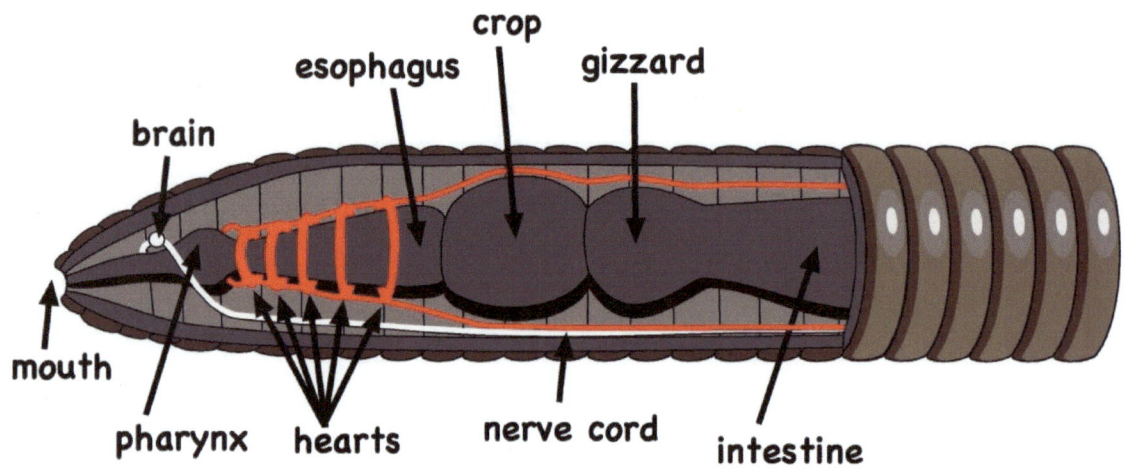

Since red wigglers don't have teeth, the gizzard grinds the food.

The food moves through the intestines, where bacteria and enzymes break it down further.

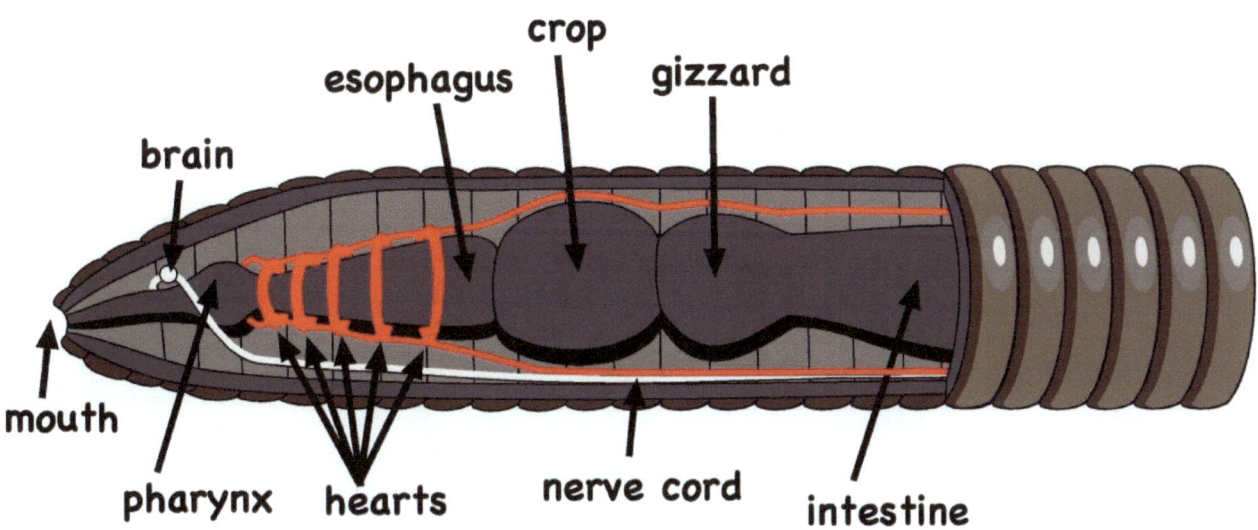

brain

esophagus

crop

gizzard

mouth

pharynx

hearts

nerve cord

intestine

The food and its nutrients are absorbed by the worm's body. The waste products (or poop) are called castings.

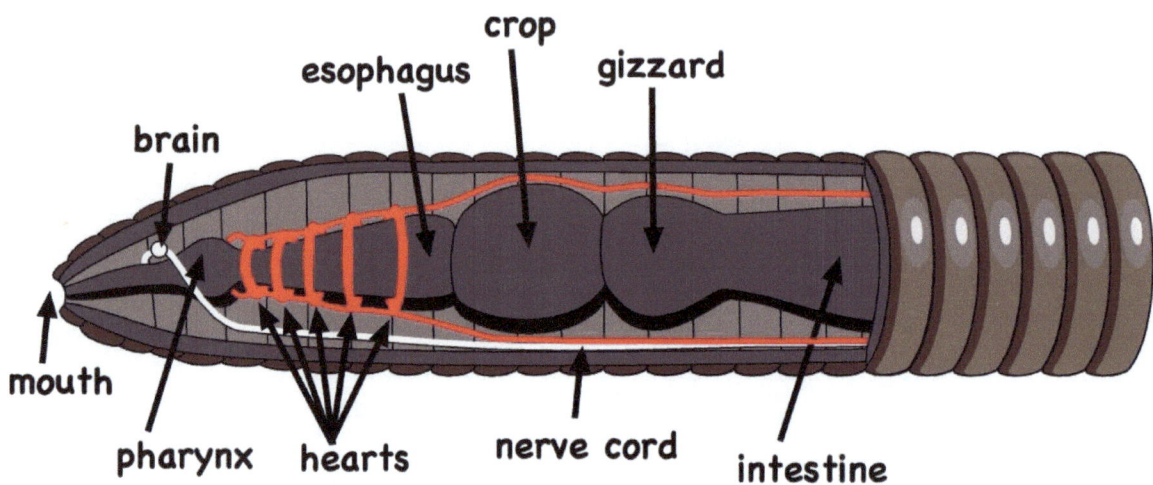

# What part of the body do the castings come from?

The castings, or poop, depart the worms' body from the anus at the tail end of the worm.

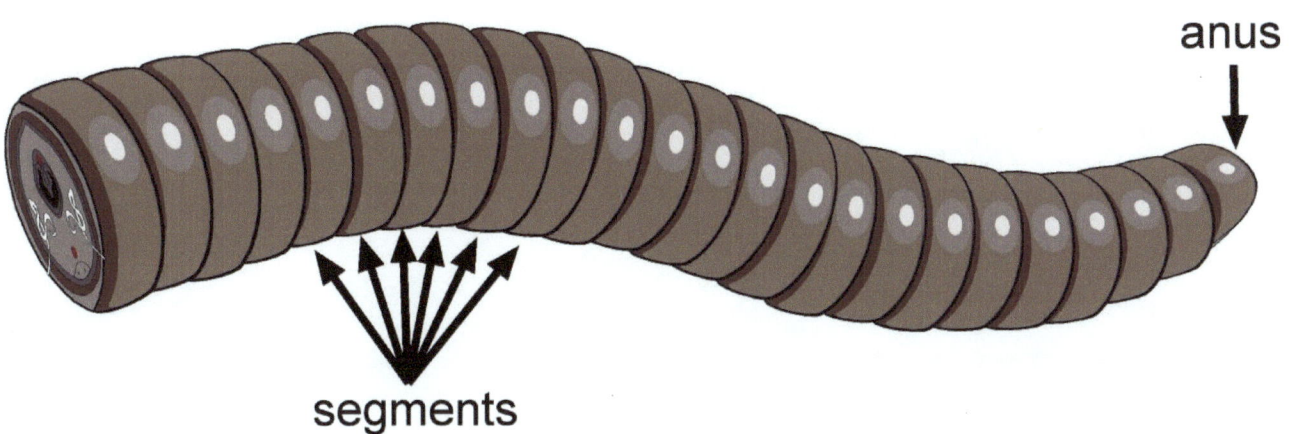

anus

segments

# What is so special about worm poop?

Worm poop is called castings.  The castings provide food that plants need to grow.

*castings*

# What happens when a worm is cut in half?

The worm's body has nerves and is sensitive. The worm has organs that it needs to live, so the worm will die if it is cut in half.

# What is slime?

Worms have special cells in their bodies that make a secretion called slime. Slime helps to keep the worms moist so they won't dry up. It also keeps their skin healthier.

# What do red wigglers eat?

You can feed your worms leftover food scraps from your kitchen. Do not put meat, cheese, or oily foods in your bin. Keep your bin wet like a damp sponge.

Worms also eat paper, cardboard, and dryer lint.

# My Journal

| Food scraps I have added to the bin | Things that move in my bin |
|---|---|
|  |  |
|  |  |
|  |  |
|  |  |

# My Journal

|  |  |
|---|---|
|  |  |
|  |  |
|  |  |
|  |  |
|  |  |
|  |  |

# My Journal

| | |
|---|---|
| | |
| | |
| | |
| | |
| | |
| | |

# How do I set up my own worm bin? Megan will show you.

The worms need air, bedding, and food to eat.

**Air:** You will need to get a bin with a lid (fifteen quarts or larger.) Have a grownup drill quarter-inch holes two inches apart, about two inches from the top of the bin.

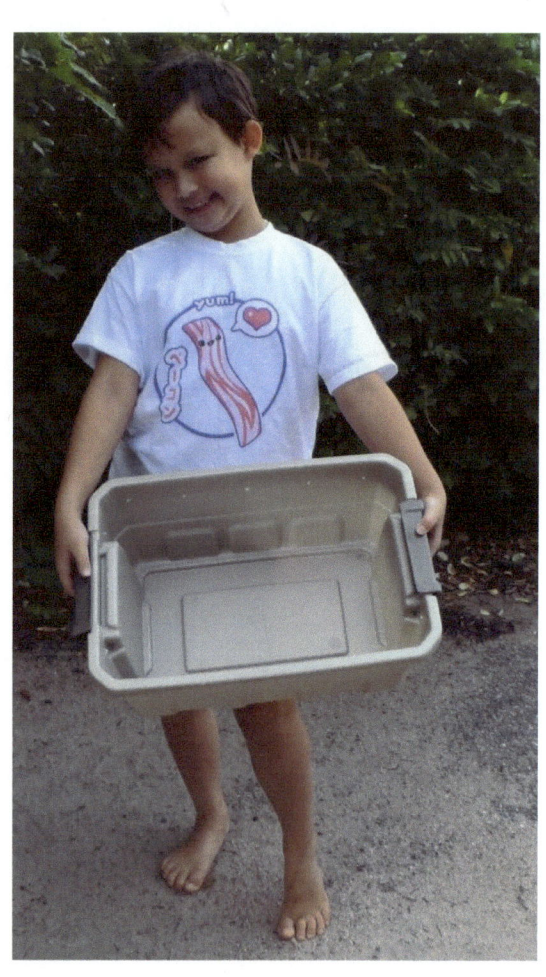

**Bedding**:  You will need to make a bed for your worms.  They need to be in wet bedding that consists of newspaper, shredded paper, cardboard, and peat moss or coir (the inside of a coconut shell).  You can add a handful of dryer lint as well.

Mix it all together.  It should be like a damp sponge-- not wet enough to drip.

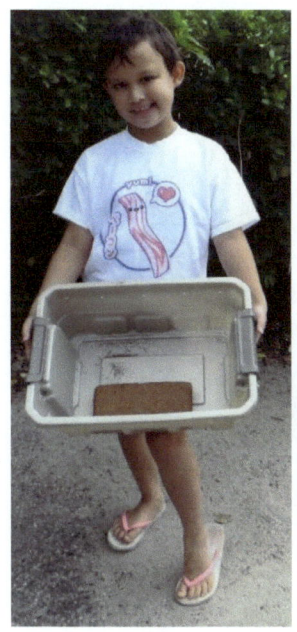   

**Food**: Add scraps of food daily—approximately half-cup a day. You can bury the food.

Your bin should consist of fifty percent fiber (paper, cardboard, peat moss, coir) and fifty percent food.

Keep the bin moist. Remember worms breathe through their skin.

Wet a piece of cardboard and place it on the top of the bedding. Add worms to your bin (under the wet sheet of cardboard.)

Place the lid on your bin.

Look for other paper and cardboard and some dryer lint to add to your bin when this paper and cardboard disappear.

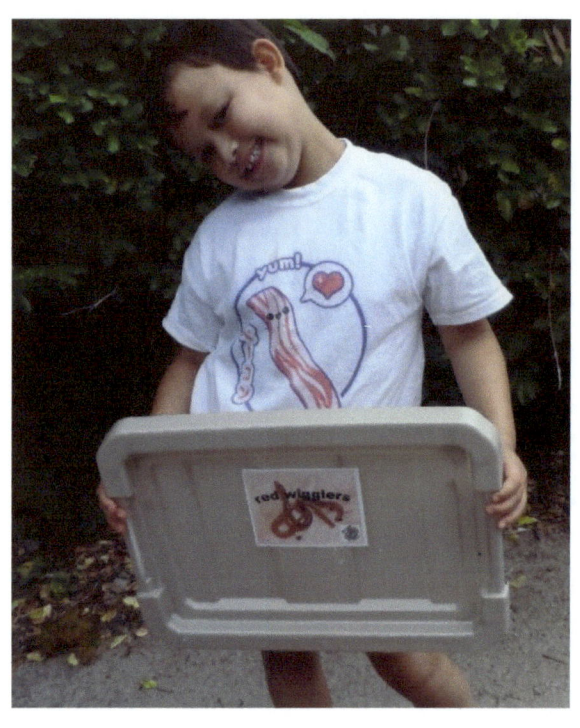

# How do I separate the worms from the castings?

After two months, it is time to separate the worms from the castings. Dump your bin on a plastic bag or in a pan and put it under a light or in the sun.

Worms don't like light.  They will keep burrowing down.

Scrape the top part off and use it for your plants. Put the worms in a separate container. Have a grownup help you.

If you have too many worms, you can share them with a friend.

# Where can I get my worms?

The Worm Ladies of Charlestown, Inc. sell worms and coir.

www.wormladies .com    401-322-7675

When you get the worms in the mail, add them to your bin.
Watch them crawl into the bedding.